# Elevated Blood Lead Levels Among Chicago Children

## Assessing Exposure and Risks for Lead Poisoning

Deidre White, M.A.

# ACKNOWLEDGMENTS

I first must thank my thesis advisor & professor Dr. Tekleab Gala who consistently steered me in the right direction. Whose door was always open whenever challenges came up with my thesis. I would also like to thank my advisor & professor Dr. Block for his commitment to his students.

I acknowledge Sargent John Cleggett of Chicago Police department for his time and assistance in identifying violent crimes within the data used for this thesis.

I give special thanks to my family and friends for the countless hours of listening to me repeatedly.

# Table of Contents


Abstract     i

List of Figures     ii

List of Tables     iii

Abbreviations     iv


**<u>Chapter 1: General Introduction</u>**     1

**1.1 Background of Lead Poisoning**     1

1.2 **Lead Poisoning in Chicago**     3

**1.3 Research Goal and Objectives**     4


**<u>Chapter 2: Literature Review</u>**     6

**2.2 Lead and Benefit**     6

**2.2 Hazards of Lead Poisoning**     7

**2.2.1 Criminal Behavior**     8

**2.2.2 Attention Deficit Hyperactivity Disorder (ADHD)**     8

**2.2.3 Social Economic Impact and Disparities**     9

**2.3 Pathways**     10

**2.3.1 Soil and Gardening**     10

**2.3.2 Employment**     11

**2.3.3. Children Hands to Mouth**     11

**2.3.4 Toys**     12

**2.3.5 Smelter**     12

**2.4   GIS Application of Elevated Lead Poisoning**                                              13

**2.4.1  Choropleth Maps**                                                                        13

**2.4. 2  Spatial Autocorrelations**                                                              14

**2.4.3  Overlay Analysis**                                                                       15

**2.4.4.  Spatial Relationship**                                                                  16

**2.4.4.1 Ordinary Least Squares (OLS) Regression**                                               16

**2.4.4. 2 Geographically Weighted Regression (GWR)**                                             17


<u>**Chapter 3: Material and Methodology**</u>                                                     19

**3.1   Overview of the Study Area**                                                              19

**3.2   Data Acquisition and Description**                                                        20

**3.2. 1  Blood Lead Levels (BLLs), Poverty Rate, and Per Capital Income**                        21

**3.2.2   Illinois Standards Achievement Test (SAT)**                                             22

**3.2.3   Violent Crime Incidences**                                                              22

**3.2.4   Percent African American Population**                                                   23

**3.2.5   Hardship Index**                                                                        24

**3.3      Research Methodology**                                                                 25

**3.3.1    Spatial Analysis Blood Lead levels (BLLs) Among the Children Aged 0-6 Years in the Neighborhoods of Chicago**                                                                        25

**3.3.2    Modeling Spatial Associations of BLL Among Children (i.e. 1-6 ages) and Associated Socio-economic, Behavioral, Cognitive Factors**                                                     27

**3.3.3  Validation Methods**                                                                     28

**<u>Chapter 4: Result and Discussion</u>**　　　　　　　　　　　30

**4.1　Blood Lead Levels (BLLs) Among the Children Aged 0-6 Years Among the Chicago Neighborhoods**　　　　　　　　　　　30

**4.2　Spatial Relationship of Elevated Lead Blood Level and Crime Occurrence Rate and Test Scores for the Chicago Neighborhoods**　　　　　　　　　　　33

**4.3　Analyzing the Spatial Relationships of Elevated Lead Blood Levels and Socioeconomic Characteristics of the Chicago Neighborhoods**　　　　　　　　　　　36


**<u>Chapter 5: Conclusions and Recommendations</u>**　　　　　　　　　　　39

**5.1　Research Findings**　　　　　　　　　　　39

**5.2　Research Significance**　　　　　　　　　　　40

**5.3　Research Limitations**　　　　　　　　　　　41


**References**　　　　　　　　　　　44

**ABSTRACT**

Lead (Pb) is a bluish-white, lustrous, soft, malleable, ductile, and poor electroconductive metal that has various uses to mankind. However, when this useful metal is absorbed by the body, various health effects are induced, often characterized as lead poisoning. In the United States, lead poisoning has decreased as a whole in recent years; however, it remains a major concern to children, still affecting 1 out of every 6 in Chicago.  Therefore, the goal of this study is to evaluate the spatial distribution of aggregated children's elevated Blood Lead Levels (BLLs) in Chicago's neighborhoods as well as its relationship with the social-economic, behavioral, and cognitive factors. Geovisualization, geospatial pattern analysis, and spatially-resolved spatial modeling tools built in ArcGIS were used. Accordingly, a significant geographical control of the BLLs was detected such that lower BLLs were detected in the central, northern, far northern, and southwestern sides of the city, while the higher BLLs were detected in the western, southern, and southwestern sides of the city (i.e., I = 0.34, permutation 999, and p-value 0.001). This distribution has shown statistically significant associations (i.e., $R^2$ = 40 – 54; and P < 0.05) with the social-economic, behavioral, and cognitive variables, indicating the likelihoods of incidences of violent crimes, poverty, and minority and lower students' performances in the higher BLLs areas. However, it is not clear if these associations imply causations to the higher/lower BLLs or vice versa. Therefore, further studies would be critical to establish how much of these associations are the causations.

Keywords: Lead poisoning, Blood Lead Level (BLL), Crime, Chicago, Children, Spatial analysis (GIS)

**LIST OF FIGURES**


Figure 1: Ripple Effects of Elevated Lead Poisoning ................................................ 11

Figure 2: The study area................................................................................................ 29

Figure 3: The 2017 data of elevated lead blood level among the children aged 0 – 6 year among community areas of Chicago ............................................................... 41

Figure 4: Geographically Weighted Spatial Relationships of elevated lead blood level and the crime occurrence and students' test scores for the 77 community areas of Chicago ........................................................................................................................ 44

Figure 5: Geographically Weighted Spatial Relationships of elevated lead blood level: a) percent African American population, b) economic hardship index, c) population's percent below poverty line, and d) household per Capital income of the community areas of Chicago.................................................................................................... 47

**LIST OF TABLES**

Table 1: Relationships of blood lead level among the children and ISAT score and

Crime Occurrence ................................................................................................. 42

Table 2: Relationships of elevated lead blood level among the children and

socioeconomic characteristics ................................................................................. 45

| | |
|---|---|
| BLLs | Blood Lead Levels |
| WHO | World Health Organization |
| GIS | Geographical Information System |
| CDC | Disease Control and Prevention |
| ACCLPP | Advisory Committee on Childhood Lead Poisoning Prevention |
| Pb | Lead |
| ADHD | Attention deficit hyperactivity disorder |
| IQ | Intelligence Quotient |
| NTP | National Toxicology Program |
| GWR | Geographical Weighted Regression |
| OLS | Ordinary Least Square |
| ISAT | Illinois Standards Achievement Test |

## 1.    Background of Lead Poisoning

Over the past 60 years, elevated BLLs have changed drastically. In 1960, the standard was set at 60 deciliters. University of Pittsburgh psychiatrist Herbert L. Needleman and others began observing "silent lead poisoning" in children with BLLs below the establishment limit (Needleman et al., 1996). These studies revealed these children had low IQ scores, attention problems, and antisocial tendencies.  As more and more reports of these deficits filtered in, the Center for Disease Control and Prevention (CDC) lowered the BLL it deemed acceptable for children further and further: In 1970, the amount was 40 micrograms per deciliter, and by 1991, it was 10 micrograms per deciliter.

The CDC recently confirms there are no safe levels of elevated lead (Meyer et al., 2003; Jones et al., 2009; Wheeler, & Brown, 2013).  As of 2012, the CDC lowered the testing standard BLL of 10 micrograms per deciliter to 5 micrograms per deciliter.  Lead poisoning occurs when lead or any of its salts are absorbed into the body (Meyer, 2003).  Damage to the body is irreversible, but treatment could prevent further damage. Once ingested or inhaled, various systems within the body are affected by lead, including the central nervous, cardiovascular, immunological, endocrine (Advisory Committee on Childhood Lead Poisoning Prevention (ACCLPP) 2012), renal, and hepatic systems (Committee on Environmental Health, 2005). Children being more susceptible to lead absorption can have a negative influence with brain development and the nervous system.  Long-term exposure in young children could result in cognitive challenges, brain damage, and behavioral

problems that often lead to criminal activities (See Fig.1) (Tarrago, & Brown, 2017).

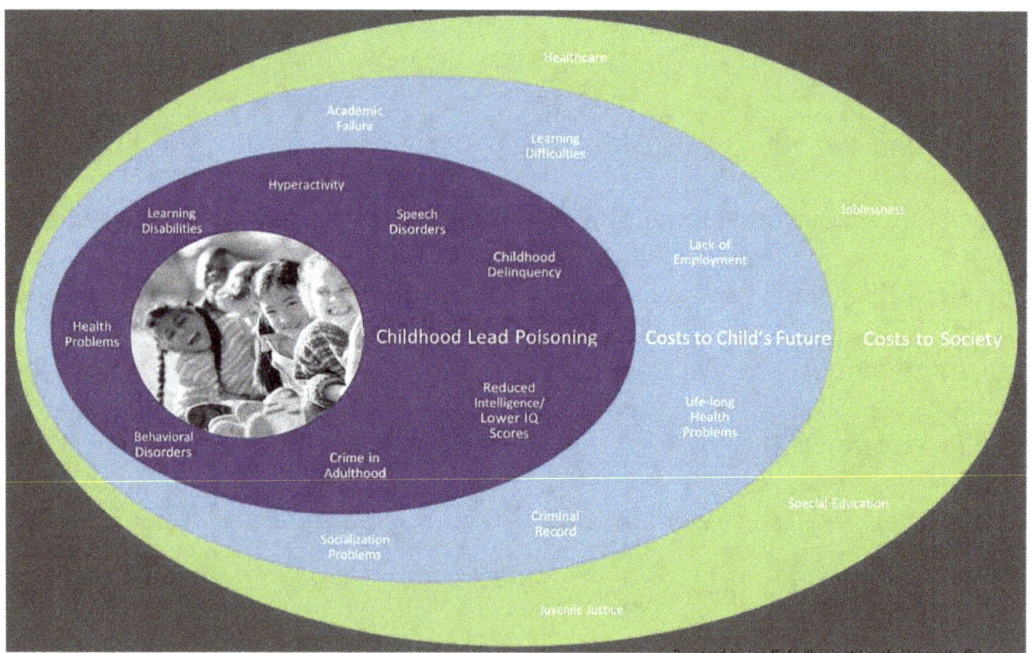


Figure 1: Ripple Effects of Elevated Lead Poisoning. (Source: Lead Safe

Illinois, Loyola University Chicago, accessed June 1, 2021)

Lead poisoning accounts for approximately 0.6% of the global burden of

disease (Sander et al., 2009; World Health Organization (WHO), 2010). In 2014 an

estimated 4 million children within the United States lived in homes that put them

at risk for cognitive and behavioral impairments due to exposure to environmental

lead (Adams, et al., 2014). Currently, pediatricians under the guidelines of the city of

Chicago are mandated to follow-up with children up to the age of 5 years old.

However, there is currently no other assistance required or provided beyond that

age.

## 1.2 Lead Poisoning in Chicago

In 2005, it was noted that Chicago lead the nation in the number of children identified with lead poisoning. Spatially dependent environmental risk factors for lead poisoning include residence or substantial time spent in a home or location built before 1978 (Kim et al. 2008). Close to 87% of Chicago homes were built before 1978, the year lead-based paint was outlawed. In a review of twenty-three research articles using geographic information systems (GIS) to model childhood lead poisoning from 1991 to 2012, the most commonly cited risk factor was age of housing (Akkus and Ozdenerol 2014). After 40 years of banning the use of lead-based paint, it continues to be a major issue in Chicago. Because the city has such a sizable number of residents, the primary concerns are to promote health, prevent disease, and reduce environmental hazards for all Chicagoans. Deteriorated paint inside houses that contained lead was believed to be the main pathway of exposure. The dynamics of long-term lead poisoning of children can affect their lives well into adulthood; moreover, one of the effects could result in criminal behavior.

Lead poisoning is an environmental hazard that can wreak havoc in young children more so than adults. By the time a child is found with an elevated BLL, the neurodevelopmental harm from the exposure may have already occurred (Sample, 2020). The elevation of lead in the blood stream became prevalent after the industrial era; to this day, prevention has become an important focus in protecting children from exposure to lead. This raises a major concern in larger cities like

Chicago.  In Chicago, over 80,000 children had elevated lead poisoning between 1996 and 2001, and even today, though the numbers are lower, childhood lead poisoning continues to be a serious problem. Although, lead poisoning rates have decreased significantly, Chicago is still one of the top cities with the largest absolute number of identified children with lead poisoning in the nation. The city of Chicago by obligation under Illinois law (Illinois lead poisoning prevention Act) mandates they conduct lead risk assessments. Therefore, this research attempts to review the geographical distribution of children tested for and having elevated BLLs, reported crimes, and schools with proven low-test scores in all seventy-seven neighborhoods of Chicago. The hypothesis is that there are viable geographical controls as well as correlations between elevated BLLs and criminal activities, low test scores, and existing socio-economic disparities in Chicago neighborhoods.

## 3.   Research Goal and Objectives

The main goal of this research is to investigate the geographical distribution of elevated BLLs among Chicago's 77 neighborhood areas, using the Exploratory spatial data analysis (ESDA) toolsets in ArcGIS. Additionally, the study deployed GIS toolsets for modeling, examining, and exploring the relationship between the elevated lead poisoning and crime, on one hand, as well as the economic status in each of Chicago's communities, on the other.

Therefore, the specific objectives are:

1. Geo-visualization and spatial dependency or nonstationary of the Blood Lead Levels (BLLs) among the children aged 0 – 6 years among the community neighborhood areas of Chicago

2. Spatial Ordinary Least Squares (OLS) and Geographically weighted (GW) regression for modeling, examining, and exploring the spatial relationship of elevated BLLs and crime occurrence rate and test scores for the Chicago neighborhoods

3. Spatial Ordinary Least Squares (OLS) and Geographically weighted (GW) regression for modeling, examining, and exploring the Spatial relationships of elevated BLLs and socioeconomic characteristics of the Chicago neighborhoods

Geographic Information Systems (GIS), has proven to be extremely beneficial in studying geographical distribution of such phenomena.  Additionally, the spatial regression modeling is established as a statistical method, and was used to examine the relationship between two spatial variables: Explanatory Variables vs. Response Variables.

**CHAPTER 2: LITERATURE REVIEW**

## 2. 1. Lead and Benefit

Lead (Pb) is a bluish-white lustrous metal with an atomic number of 82, one of the 118 chemical elements.  It is considered a heavy metal because of its density as compared with other common materials. Native lead is rare in nature and usually found in ore with zinc, silver, and copper (Lenntech, 2018). Its extraction also contains these elements. Australia is among the major lead producers of the world, accounting for 19% of its production, followed by the USA, China, Peru, and Canada. The world Lead production amounts to 6 million tons a year, and there are estimated reserves of 85 million tons worldwide, which is less than a 15-year supply.  Physically and chemically, Lead has properties of softness, high malleability, ductility, and relatively poor electric conduction. These properties provide various uses to mankind (Britannica, T. Editors of Encyclopaedia, 2020).

Lead was previously used for making pipes before it was officially banned due to its adverse health effects. Still, Lead pipes bearing the insignia are used as drains from the baths. Other Lead material such as Tetraethyl lead (PbEt4) was also used in some grades of petrol (gasoline) before it was discarded on the ground of its environmental hazards (Lenntech, 2018). Lead is a major constituent of lead acid battery, and is applied as a coloring element in ceramic glazes, electrodes in the process of electrolysis, and in the monitors of computer and television screens to shield viewers from radiation. Lead can has also been used in sheeting, cables, crystal glassware, ammunitions, bearings and weight in sport equipment (Lenntech, 2018).

## 2.2. Hazards of lead poisoning

Lead, which enters the human body mainly through eating foods, drinking water, and breathing air, is one out of four metals that have the most damaging effects on human health. The amount of Lead that enters through food accounts for 65% of the total, followed by water, which is 20% and air, which is 15% (Lenntech, 2018). Neuropsychological and biological studies find that sufficient exposure to lead is associated with brain dysfunction (Sanders et al. 2009). Scientist have found that lead exposure alters neurotransmitter and hormonal systems in ways that may induce aggressive and violent behavior (Needleman et al. 1996).  Childhood lead poisoning, both acute and chronic, remains an enormous problem.  Lead poisoning accounts for approximately 0.6% of the global burden of disease (Sanders et al. 2009).  According to published information from the Mayo Clinic, symptoms can be hard to detect initially, and even people who seem healthy can have high blood levels (Mayo Clinic, 2016).  Once the body has accumulated dangerous amounts, signs and symptoms will start to appear.

In general, lead impacts on human health can be summarized by the following symptoms: disruption of the biosynthesis of haemoglobin and anaemia, a rise in blood pressure, kidney damage, miscarriages and subtle abortions, and disruption of the nervous systems. It can also cause brain damage, declined fertility of men through sperm damage, diminished learning abilities and behavioral disruptions of children, such as aggression, impulsive behavior, and hyperactivity (Lenntech, 2018). Lead can also enter an unborn fetus through their mother's placenta and cause damage to the nervous system (Lenntech, 2018).

### 2.2.1. Criminal Behavior

Studies have linked elevated lead level with criminal behaviors (Stretesky and Lynch 2004; Knapp, 2013; Winter & Sampson, 2017). According to a study conducted by Knapp, (2013), by examining the hair of inmates convicted of violent and property crimes, they attempted to assess whether these behaviors were linked to variations in exposure to toxic substances, including lead. Additionally, Stretesky and Lynch (2004) found that, despite the inmate's age, socioeconomic status, months institutionalized, and drug use history, hair-lead levels distinguished that, of the two groups, violent offenders specifically were more likely than property offenders to have elevated levels of hair-lead (Stretesky and Lynch 2004).  Both studies established links between elevated BLLs and violent crimes for Chicago communities.

### 2.2.2. Attention Deficit Hyperactivity Disorder (ADHD)

The immediate health effect of concern in children is typically neurological. It is important to remember that childhood lead poisoning can lead to health effects later in life, including ADHD, delayed learning, and lower IQ. These effects can later impact their school performance (Tarrago, & Brown, 2017; Geier et al, 2017). Data from the National Toxicology Program (NTP) 2012 showed that the effect of concurrent BLLs on IQ may be greater than currently believed. Lead inhibits the bodies of growing children through the absorption of iron, zinc, and calcium— minerals essential to proper brain and nerve development (Needleman, & Bellinger, 1991).

### 2.2.3. Social Economic Impact and Disparities

In the past, studies conducted by Needleman et al., (1996) have shown some adverse Health outcomes are directly impacted by income levels. Where poverty exists, disparities will also. This study looked at one important environmental hazard: lead.  It also attempted to identify some major concerns surrounding its long-term effects and, through the use of Chicago maps, displaying which areas are mostly affected in correlation with the economic statuses of those areas. There is a greater likelihood of exposure in troubled neighborhoods, and life course implications produce barriers for upward mobility (Reed, 2018). There is also evidence suggesting that when minorities and the poor are not disproportionately exposed to lead in the environment, they are still more likely than whites and the affluent to suffer from the ill effects of lead (Reed, 2018).  This is because the poor and minorities have fewer contacts with physicians, and, when they do receive treatment, they are more likely than the affluent to receive treatment that is inadequate or incomplete.

Specifically looking at the case of lead poisoning, the resource-deprived are less likely than those with resources to be screened and effectively treated for lead poisoning (Reed, 2018).  Moreover, the more affluent counties do a better job at screening and treating lead poisoning than deprived neighborhoods.  In addition, poor diet and low psychological well-being, which are also associated with economic deprivation, are two conditions thought to exacerbate the effects of lead poisoning (Reed, 2018).

## 2.3. Pathways

### 2.3.1. Soil and Gardening

Lead in soil is increasingly considered a significant source of exposure, especially for children engaging in hand-to-mouth contact through early methods of mobility like crawling (Mielke and Reagan 1998). The positive relationship between soil lead values and elevated blood level has been identified by several researchers (Meilke et al. 1999; Johnson, & Bretsch, 2002).  Studies demonstrating that BLLs in children might be more related to this source than age or housing (Lanphear et al. 1998; Mielke and Reagan 1998). Meilke and Reagan (1998) found that at the census tract level, soil lead is significantly associated with blood lead in New Orleans, Louisiana. Spatial analysis of property-level soil lead tests has led to an understanding of where lead exposure is greatest: nearer the foundation of the home (termed the dripline) and the region closest to the street (Mielke 1994). The spatial distribution of soil lead concentrations has been mapped for several cities including Baltimore, Maryland (Mielke et al. 1983); New Orleans, Louisiana (Mielke et al. 2013); Syracuse, New York (Johnson and Bretsch 2002); Indianapolis, Indiana (Laidlaw et al., 2012); and Toledo, Ohio (Stewart et al. 2014). Results of these studies demonstrate that soil lead concentrates in the center of cities (Mielke et al. 1983) and in older neighborhoods (Levin et al. 2008; Mielke, et al., 2011). For example, a profile of soil lead in New Orleans demonstrated not only the variability in soil lead from an urban center to the suburbs, but also how the pattern of soil lead varied on individual properties along that continuum (Mielke 1994).

### 2.3.2. Employment

Employment is rarely considered when thinking about lead poisoning, unless a person works in a hazardous job industry. There are service industry people that are directly affected, such as: painters, battery workers, maintenance workers, construction workers, and carpenters just to name a few.  Lead inhalation, the most common exposure route for construction employees, occurs from breathing in dust or fumes that contain lead.  Once lead is inhaled into the lungs, it goes quickly to other parts of the body in the blood (Occupational Safety and Health Administration, 2007). These workers are potentially exposed daily at least eight hours a day. And they aren't the only ones at a constant risk. Consider a person's hobby is painting houses. They are just as affected by lead as those in the aforementioned professions (Tarrago, & Brown, 2017).  The plumbing industry has been replacing the old plumbing in homes with plastic instead of the galvanized steel piping previously used. This will assist in keeping our water from being contaminated.

### 2.3.3. Children Hands to Mouth

The most common way children become contaminated with lead is by hand-to-mouth contact. They are more likely to place their hands and other objects that may carry lead dust into their mouth (Ko et al. 2007; Witzling, et al., 2010). A study was done via video assessments of children's touching and mouthing behaviors during outdoor play in urban residential yards.  The assessment consisted of thirty-seven children. Their ages ranged from 1-5 years and they were observed for two hours.  The study focused on hand touches to the ground or other walking level surfaces and the respective oral behaviors (Ko et al. 2007; Witzling, et al., 2010).

The conclusion of the test showed blood lead was directly correlated with log-transformed rates of hand-in-mouth (Pearson's correlation, r=0.564, n=22, P=0.006) and object-in-mouth (Pearson's correlation, r=0.482, n=22, P=0.023) behaviors (Ko et al. 2007).

### 2.3.4. Toys

It was not until 2006 when the U.S. Consumer Products Safety Commission announced a number of toys imported from China were contaminated with lead paint. In 2007, more than 1.5 million "Thomas & Friends" wooden railway toys were reportedly painted with lead.  Mattel, one of the largest toy companies, recalled over nine hundred thousand toys after discovering they were contaminated with lead (Danger, 2008).  Even by 2008, the magnitude of this problem had not been fully determined and, hence, adequate methods for controlling the importation of contaminated toys had not yet been established.

### 2.3.5. Smelter

Smelters are contaminated sites that were once metal industrial industries and are now closed. The Environmental Protection Agency (EPA) listed this type of site, as well as other contaminated industrial sites, as Superfund sites until 1980 (Switzer, & Bulan, 2002). These contaminated sites exist nationally due to hazardous waste being dumped, left out in the open, or otherwise improperly managed. These sites include manufacturing facilities, processing plants, landfills, and mining sites (Switzer, & Bulan, 2002).

## 2.4. GIS application of elevated lead poising

Various studies have used Geographical Information Systems (GIS) to assess the geographical distribution of elevated BLLs and their relationship to other risk factors (Akkus, & Ozdenerol, 2014; Mielke et. al., 2011; Kunene, et al., 2017; Vasovic, 2020). The studies used overlay analysis of choropleth maps, Spatial autocorrelations, spatial relationships, and geographically weighted regressions (GWR).

### 2.4.1. Choropleth Maps

The choropleth maps are GIS mapping applied to quantitative data. The map uses spatially aggregated data (e.g., blood lead level (micrograms of lead per deciliter of blood, mg/dL) among children) in a given area, and displays them with various symbology (shed scales or colors) to indicate varying values. Various studies have used Choropleth maps to investigate or analyze the spatial distribution of lead hazards in the human environment (Clune et al., 2011; Mateo-Tomás et al., 2016; Peng et al., 2019). Clune et al., (2011) produced a global map of environmental lead poisoning in children as an exercise to show exposure hotspots in regions around the world, particularly where health is poor. Additionally, Mateo-Tomás et al., (2016) adopted choropleth maps to demonstrate the Spatio-temporal varying areas of risk to lead exposure. Moreover, Peng et al., (2019) used Choropleth maps to investigate the spatial distribution of lead contamination in soil and equipment dust at children's playgrounds in Beijing, China.

### 2.4.2. Spatial autocorrelations

Spatial autocorrelation is a geostatistical tool of GIS used to depict if there is a presence of statistically significant spatial patterns on the values of the variable under consideration (e.g., blood lead levels (micrograms of lead per deciliter of blood, mg/dL) among children). A positive spatial autocorrelation denotes the spatial patterns in which variables with similar values are geographically congregated, while a negative spatial autocorrelation denotes the patterns where variables with dissimilar values are neighboring each other.  A zero spatial autocorrelation denotes the situation where values of the variable have no pattern or are randomly distributed.  The spatial autocorrelation uses global or local Moran I indices to assess these patterns in the data.

Several studies have used spatial autocorrelation to investigate patterns in the geographical distribution of elevated blood lead levels among children or other health occurrences (e.g., Akkus, 2016; Kunene et al 2017; Xiao et al., 2018). Akkus, (2016) spatial inquiry into childhood lead poising in Shelby County, TN and used the Global (Moran's I) and local spatial autocorrelations (Getis and Ord's Gi), to find geographic controls as well as local pockets of high at-risk areas. On the other hand, Kunene et al., (2017) also deployed both global as well as location spatial autocorrelation to understand the spatial pattern of the adult HIV/AIDS prevalence rates and the risk factors in Sub-Saharan African countries. Similarly, Xiao et al., 2018 used spatial autocorrelation for monitoring patterns in the data of heavy metals (cadmium, lead, total arsenic, total chromium, and total mercury) in China found varying levels of geographical controls on heavy metal contamination of the rice field.

### 2.4.3. Overlay Analysis

Overlay analysis is a GIS operation that allows the comparisons of the attribute and location information about multilayered GIS data representing different themes. The analysis-which may include the operations such as but not limited to, a spatial union, intersect, and subtract—is used for answering questions related to spatial relationship, suitability modeling, and optimal site selection. Various studies have used the overlay analysis of the spatial relationship of childhood lead poisoning and other socio-economic and environmental factors (Vaidyanathan et al., 2009; Miranda et al., 2011; Mielke et al., 2013). Reissman et al., (2001) established a strong relationship between the old housing and blood lead levels, with overlay analysis, in Jefferson County, Kentucky, USA. Additionally, Vaidyanathan et al., (2009) assessed the spatial relationship of the neighborhood at risk for children subjected to lead exposure and neighborhood-level blood lead testing of young children living in the city of Atlanta, Georgia, and found a strong correlation. Moreover, Miranda et al., (2011) used a spatial overlay for analyzing the relationship of the local airports and blood lead levels and found a significant positive association with children living in six counties in North Carolina. Finally, Mielke et al., (2013) deployed the GIS-based overlay analysis to evaluate the spatial correlation between the New Orleans soil Pb and Children's blood Pb and found a statistically significant relationship.

### 2.4.4. Spatial relationship

Spatial regression is a GIS technique of modeling, examining, and exploring relationships between variables designated as dependent and independent variables. In general, the spatial regression, however, does not indicate causation between variables and therefore does not serve for prediction purposes. However, it is used to describe the strength/weakness of the relationships, the direction of the relationships (positive, negative, and zero), and the significance of the goodness of fit between two variables.

Spatial data often do not fit the traditional assumptions of non-spatial regression. These are normality (i.e., that for any fixed value of the independent variable, the values of the dependent variable are normally distributed), homoscedasticity (i.e., the variance of residual error is the same for any value of the independent variable), and independence (i.e., observations are independent of each other). Instead, spatial data are autocorrelated (i.e., the values of variables that are near each other are more similar than those values of variables further away) and nonstationary (i.e., the spatial patterns of the values of variables vary based on their geographical location). The spatial lag model and spatial error model are introduced into the spatial regression to account for these properties of spatial data. Two types of spatial regression models, Ordinary Least Squares (OLS) regression and Geographically weighted regression (GWR), are used.

### 2.4.4.1. Ordinary Least Squares (OLS) regression

Ordinary Least Squares (OLS) regression is a well-known simple regression technique and is often used as a good starting point for all spatial regression analysis. OLS is a global spatial regression model where a single equation of models examines and explores the relationships in the entire dataset. Several studies have used OLS for evaluating special relationships of blood lead levels among children and its determinants (e.g., Boutwell et al.,2017; Fokum et al., 2017). For instance, Boutwell et al., (2017) established the spatial association of aggregated blood lead levels and specific indicators such as gun violence, homicide, and rape. The spatial association was established using the spatial generalized linear mix model (SGLMM). The model accounts for spatial dependent data not following Gaussian distribution and found that aggregated blood lead level is a significant predictor of violent crimes at the geographical scale of census tracts for the city of St. Louis MO. Additionally, Fokum et al., (2017) estimated the prevalence and exposure of children to lead poisoning in Illinois. Illinois law requires health providers to obtain a blood lead test and assessment at least at the age of one and two years. Such tests and assessments are required as evidence before a child attends licensed daycare, kindergarten, or school. And this blood lead level among children is used to analyze the prevalence of childhood lead poisoning in Illinois.

**2.4.4.2. Geographically weighted regression (GWR)**

Geographically weighted regression (GWR) is a spatial regression technique, whose utility for spatial has increased in recent years. It provides a local model of spatially varying relationships among variables by fitting a regression equation to every location in the dataset. And therefore, GWR is a more reliable and powerful model of examining and estimating spatial linear relationships, should it be used properly. Although GWR application to blood lead levels and its determinants is limited, several studies have used it for modeling spatial-resolved relationships in the dataset concerning social and health services (e.g., Vasovic, 2020; Holmes, 2021). To investigate the controls of spatial disparity of unemployment in Chicago, Vasovic (2020) used GWR, to model spatially varying relationships of neighborhood race/crime and unemployment.  Similarly, Holmes, (2021) used GWR to examine settlement patterns of older African Americans in Chicagoland and relations with the patterns of African Americans of all ages and older Americans of all races in Chicagoland.

**Chapter 3: MATERIAL AND METHODOLOGY**

## 3.1. Overview of the Study Area

The city of Chicago covers just over two hundred and thirty-one square miles, approximately five hundred and seventy feet above sea level. It is situated along the shore of Lake Michigan the 5th largest body of fresh water in the world. The city was incorporated as a city in 1837 and currently has a population that amounts to over two million people and has the third largest school system in the United States, educating over three hundred and fifty thousand students a year.

Chicago is sectioned into seventy-seven different neighborhood community areas. The community areas of Chicago were defined by University of Chicago's social science research committee for the city's statistical and planning purposes. Neighborhood community areas are a geographical scale at which various socio-economic, demographic, and environmental statistical data are traditionally aggregated and widely used by the city to assist the local and regional urban planning. The city is also divided into 9 districts: central, north, far north, northwest, west, southwest, far southwest, south, and far south sides. These districts are generally classified along Chicago's racially segregated fault lines.

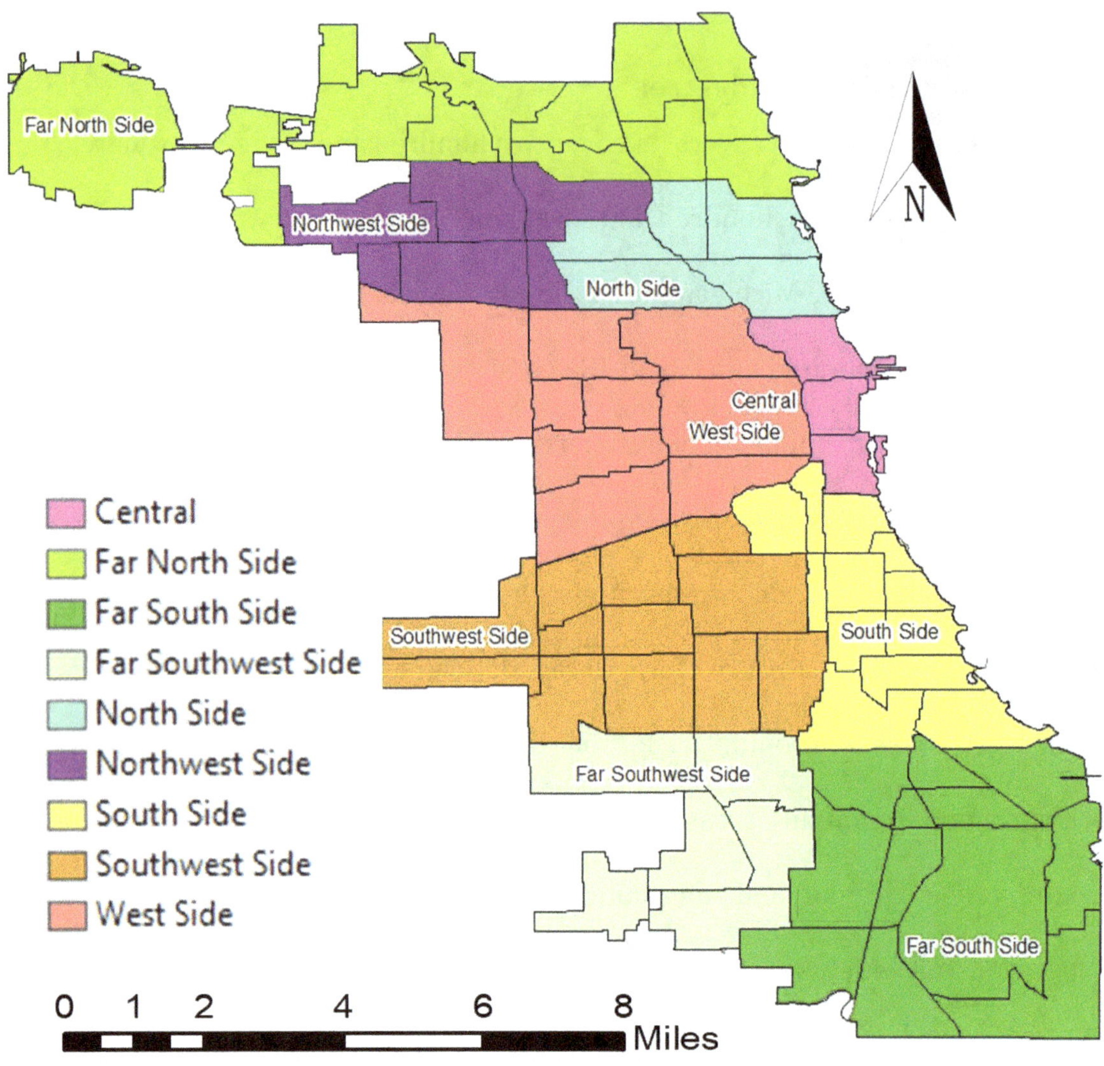



Figure 2: The study area

## 3.2. Data Acquisition and Description

This study deployed the data of blood lead level among children aged 0 – 6 in Chicago and compared it to students' performances on state wide test scores and crime occurrence rates. Additionally, it used data of a percentage of the African American population, people below poverty level, per capital incomes, and values of Hardship index. This data aggregated to conduct the analysis at the scale of neighborhood community areas in Chicago.  Although the data analyses were conducted at the scale of Chicago neighborhoods, the results were presented and interpreted at the scale of these districts so as to discuss the BLLs values among Chicago children and associated social-economic, behavioral, and cognitive factors in the context of socio-environmental justice.

### 3.2.1. Blood lead levels (BLLs), Poverty rate and Per Capital Income

The blood lead levels (BLLs), people below poverty level, and per capital incomes were obtained from Chicago Health Atlas (https://www.chicagohealthatlas.org/). The BLLs, which measured in micrograms of lead per deciliter of blood (mg/dL), is listed under the category of the environmental health. It is a measurement of an elevated blood lead level among the children aged 0 – 6 in Chicago, with corresponding 95% confidence intervals, by Chicago community area, for 2017. On the other hand, the per capital income and poverty rates are listed under the income category of the socio-economic factors and are found aggregated at the scale of neighborhood community areas of Chicago.

The Per capital income is the average income earned by individuals, whereas poverty rate is the ratio of the number of people whose income falls below the poverty line. The poverty line is a threshold nationally defined as when a household income for a family of four earns less than $26,200 per annum. In general, the city of Chicago and the Chicago Department of Health together prepared the Chicago Health Atlas data for better understanding and monitoring, and for solutions for improved Chicago health and well-being.

### 3.2.2. Illinois Standards Achievement Test (ISAT)

The data of Illinois Standards Achievement Test (ISAT) for the city of Chicago was obtained from Chicago public schools (https://www.cps.edu/about/district-data/metrics/assessment-reports/ ). The ISAT data is used for assessing the learning standards of the Illinois students and aggregated by schools in the city. It consists of the students' assessment of math, reading, and writing tests, and reported the percentage of students who are exceeding the standards needed. It is expressed in the percentile ranks, which range from as low as 1 to the maximum of 99%, with 50% denoting a school's average performance. The school aggregated performance ranks of the 2013/14 school year for math, reading, and writing tests were averaged to get estimates of the overall ISAT scores of schools in the city. The school performances were averaged across the community areas in preparation for further analysis.

### 3.2.3. Violent crime Incidences

Data for crime incidence report is created by the research & development division of the Chicago Police and was obtained from the city of Chicago data portal (https://data.cityofchicago.org/). The crime data is prepared by Chicago police department's Citizen Law Enforcement Analysis and Reporting (CLEAR) system. In total, a spreadsheet containing incidences of 265,268 crimes was downloaded of which 56,725 were violent crimes. Violent crimes are, generally, offenses committed against a person. According to the state of Illinois, violent crimes include incidences of battery, robbery, burglary, arson, child abuse and endangerment, homicide, kidnapping, rape, assaults, and offenses. Violent crime data conatins coordinate points for the locations where the crimes were committed.  The add XY event data tool of ArcGIS is used for geocoding the point data and mapping. Finally, the point-based violent crime data was joined with the base-map of neighborhood community area map, by using spatial join, and presented a spatially-resolved summary of the number of crimes committed per each neighborhood in the city.

### 3.2.4. Percent African American Population

The percent African American population data of 2017 was obtained from the US census bureau data portal (https://www.census.gov/programs-surveys/acs/data.html). The data is collected by the American Community Survey (ACS), which gathers demographic data used for monitoring and conducting public services planning. The data is available at many geographic aggregate levels depending on the survey frequency. This study used 5-years of ACS survey data, which presents demographic (i.e., ancestry, citizenship, educational attainment, income, language proficiency, migration, disability, employment, and housing characteristic) data at a scale of census tract. Parallel census tracts boundary data of Chicago, to which the tabular demographic data is joined, was obtained from Topologically Integrated Geographic Encoding and Referencing (TIGER) Line Shapefiles data (https://www.census.gov/geographies/mapping-files.html). The two data are joined using common primary keys of the attribute tables.  Once the demographic data is joined, the population data was aggregated to the scale of Chicago community neighborhood area. Percent African American population is a quotient of African American population to the total population of the community areas.

### 3.2.5. Hardship Index

The hardship index is a measure that is used for comparing the socio-economic conditions by combining six social and economic variables (Nathan, & Adams, 1976; Wilson, et al., 2017). These are unemployment rate (i.e., percent of unemployed population over 16 years of age), educational achievement (ratio of population ages 25 or above without high school diploma), and per capita income (i.e., individual income per annum). It also includes dependency ratio (the proportion of population below age 18 and above 64), crowded housing (percentage of the housing units where more than a person live in a room), and poverty (i.e., proportion of the population below designated state's poverty level).  The 2017 ACS data of these variables aggregated at the scale of neighborhood community areas of Chicago were obtained from the Chicago Health Atlas (https://www.chicagohealthatlas.org/). The hardship index is calculated by standardizing the socio-economic variables (e.g., unemployment rate) using the following equation.

$$X = ((Y\text{-}Y_{min}) / (Y_{max}\text{—}Y_{min})) * 100$$

Where, X = standardized unemployment rate of the neighborhood community areas of Chicago

Y = the unemployment rate of the neighborhood community areas

$Y_{min}$ = the value of the minimum unemployment rate of all neighborhood community areas of Chicago

$Y_{max}$ = the value of the maximum unemployment rate of all neighborhood community areas of Chicago

The final hardship index is the average of the standard values of the six socio-economic variables considered. The index has a value ranging between 0 and 100. Zero represents neighborhoods experiencing the least socioeconomic hardship, while 100 represents neighborhoods experiencing with a stressful socioeconomic hardship.

## 3.3. Research Methodology

### 3.3.1. Spatial Analysis Blood Lead Levels (BLLs) among the children aged 0 – 6 year in the neighborhoods of Chicago

The spatial analysis of the BLLs of the children in the urban neighborhood of Chicago was conducted via geovisual and geostatistical spatial pattern analysis (Kunene, 2016; Kunene, et al., 2018; Hunley III, 2019).  The geovisual analysis works by visually interpreting the spatial distribution of the values of aggregated BLLs (i.e., deciliter of blood (mg/dL)) after the data that is being displayed as a choropleth map on the Arcmap. The interpretation was done in such a way that the data is classified into five categories around the city's, state's, and national average.

On the other hand, the geostatistical spatial pattern analysis was conducted using global as well as local Moran's I statistics, both built in a spatial statistics toolbox in ArcGIS. The Global Moran's I is a statistical tool used for measuring the tendency that proximate urban neighborhoods are having similar, dissimilar, and random values of the BLLs. The statistics of Global Moran's I index (Anselin, 1995; Hunley, & Gala, 2020) is estimated by:

$$I = \frac{\sum_{i=1}^{n}\sum_{j=1}^{n} W_{i,j}[X_i - \overline{X}][X_j - \overline{X}]}{\left(\sum_{i=1}^{n}\sum_{j=1}^{n} W_{i,j}\right)\sum_{i=1}^{n}(X_i - \overline{X})^2} \qquad (1)$$

Where: I is the Moran's I statistics, X is the values of the BLLs among the children of urban neighborhoods i, $X_j$ are the values of the BLLs among the children of urban neighborhoods j, $W_{i,j}$ are the spatial weights that determines the relationship between urban neighborhoods i and j, and n is equal to the total number of urban neighborhoods. The statistical significance of the spatial pattern is established by comparing the observed Moran's I statistics with expected Moran's I, which is random distribution. The Expected Moran's I ( is given by:

$$E(I) = \frac{-1}{n-1} \qquad (2)$$

The local Moran's I statistics, conversely, is a measure of local hot or cold spots of the BLLs values of the children among Chicago neighborhoods. Hotspots are areas where clusters of higher values of BLLs are detected, while coldspots are localities where clusters of lower BLLs values are detected. The statistical tool deployed to measure the local Moran's $I_i$ statistics is Anselin Local Moran's $(I_i)$ (Anselin, 1995; Mitchell, 2005; Hunley, & Gala, 2020), and is expressed as:

$$I_i = \frac{X_i - \bar{X}}{S_i^2} \sum_{j=1}^{n} W_{i,j} (x_i - \bar{X}) \qquad (3)$$

Where: $X_i$ is values of the BLLs among the children for urban neighborhoods, i is the mean of the aggregated values of the BLLs among the children of Chicago city, and $W_{i,j}$, is spatial weight between neighborhoods i and j and is a standard deviation of the values of the neighborhoods BLLs given by:

$$S_i^2 = \frac{\sum_{j=1}^{n}(X_j - \bar{X})^2}{n-1} \qquad (4)$$

The standard deviation is a measure of the standardized variation of the values of the BLLs among the children for urban neighborhoods from its mean.

### 3.3.2. Modelling Spatial Associations of BLL among Children (i.e., 1 – 6 ages) and associated social-economic, behavioral, cognitive factors

The spatial associations of the BLLs and risk factors were conducted globally using a simple spatial regression model as well as locally by deploying a Geographical Weighted Regression (GWR) model (Kunene, 2016; Kunene, et al., 2018). The simple spatial regression is an extension of a regular Ordinary Least Square (OLS) regression, which involved the creation of spatial weight structure. The spatial weight defined the structure to account for the non-stationarity in the data, i.e. a situation where mean and variance are dependent on the location and distance between neighborhoods.  Finally, the global model predicted neighborhoods' aggregated BLL among children of Chicago as influenced by values of social-economic, behavioral, and cognitive factors.

On the other hand, geographically weighted regression (GWR) establishes local regression that would help predict values of BLL as a condition by each associated factor.  GWR works based on a simple idea that models of local goodness-of-fit are developed from subsets of values of the BLLs and factors scanned with a moving window. The number of neighborhoods considered for the estimation is contingent on the size of the moving window (i.e., Kernel type, Bandwidth method, Distance, and Number of neighbors' parameters). This model also accounts a nonstationary while estimating the unbiased slope, and intercept for the goodness-of-fit. The equation to estimate the dependent BLL among children $Vj$ from set of n social-economic, behavioral, and cognitive factors $Z_i$, for a neighborhood pi is locally given by Böhner, & Bechtel, (2018) as follows:

$$\hat{V}_j = a_j + \beta_j z_j + \epsilon_j \qquad (5)$$

$$a_j = \frac{\sum_{i=i}^{n} w_{ij} z_i^2 \sum_{i=1}^{n} w_{ij} v_i - \sum_{i=1}^{n} w_{ij} \sum_{i=1}^{n} w_{ij} z_i v_i}{\sum_{i=1}^{n} w_{ij} \sum_{i=1}^{n} w_{ij} z_i^2 - \left(\sum_{1=i}^{n} w_{ij} z_i\right)^2} \qquad (6)$$

$$\beta_j = \frac{\sum_{i=i}^{n} w_{ij} \sum_{i=1}^{n} w_{ij} z_i v_i - \sum_{i=1}^{n} w_{ij} z_i \sum_{i=1}^{n} w_{ij} v_i}{\sum_{i=1}^{n} w_{ij} \sum_{i=1}^{n} w_{ij} z_i^2 - \left(\sum_{1=i}^{n} w_{ij} z_i\right)^2} \qquad (7)$$

Where: $a_j$ is the intercept, $_j$ is the beta coefficient, and $\varepsilon_j$ is the error term of the regression equation. The observed values $V_i$ are weighted by $W_{ij}$ indicating the spatial weights for the input data of the target neighborhood $P_i$.

Simply, spatial weights are often estimated by giving a weight of 1 for the neighborhoods that fall within the spatial kernel and 0 for those outside.  However, in this study, the adaptive kernel, i.e. Gaussian weighting scheme, the weight assigned to the neighbor decreases when distances are applied. The adaptive kernel (Gaussian) is given by:

$$W_{i,j} = e^{\frac{1}{2}\left(\frac{d_{i,j}}{b}\right)^2} \qquad (8)$$

Where: $W_{ij}$ are weights of the neighborhood areas in Chicago with respect to the target neighborhood $P_i$, and $d_{i,j}$ are neighborhoods distances from the target, and b is "bandwidth".

### 3.3.3. Validation Methods

The Spatial Autocorrelation and Hot Spot Analysis (i.e., Getis-Ord Gi* statistics) of the BLLs among the children were verified by the values of the Moran's I Index and corresponding p-value. Generally, while +1 Moran's Index value indicates spatial clustering of the aggregated data of BLLs value, the index value near -1.0 indicates the spatial dispersion, and the 0-index value depicts spatial randomness. P-values (i.e., $\alpha < 0.05$) ascertain the statistical significance of the observed patterns of dispersion or cluster in the distribution of BLLs among the neighborhoods of Chicago.

On the other hands, various indices were deployed to evaluate performances of the Simple spatial and geographically weighted regressions. The regression models were implemented to establish association between BLLs among the children and associated social-economic, behavioral, and cognitive factors. Generally, the indices validated the associations in terms of strengths, directions and their statistical significances. While the relative strengths of the associations were evaluated with coefficient of determination ($R^2$), the directions were validated by the positivity or negativity of the $\beta$-coefficient and statistical significances denoted against by the P-values (i.e., $\alpha < 0.05$).

## 4.1. Blood Lead Levels (BLLs) among the children aged 0 – 6 year among the Chicago neighborhoods

Figure 3 evaluated the spatial patterns and distribution (a), geographic controls of these spatial patterns and distribution (b), and spatial clustering of similar values of elevated BLLs (c). Accordingly, the BLLs among children aged 0 – 6 year for Chicago ranges between 0 – 7.90 micrograms of lead per deciliter of blood (mg/dL). The citywide average BLLs among children is $\mu$ = 1.9 mg/dL ± $\sigma$ = 1.85. According to Illinois Department of Public Health (2007), this is higher than the average national (i.e., 1.6 mg/dL) as well as the Illinois State's (i.e., (i.e., 1.3 mg/dL) BLLs among children age 0 – 6.

The spatial patterns of the BLLs among children is distributed in such a way that lower levels are found in the central, northern, far northern, and southwestern sides of the city while the higher values are found in the western, southern and southwestern sides of the city. Further analysis was conducted to see if there are significant overall geographical controls within the spatial patterns of the BLLs among children in Chicago (i.e., Figure 3b). Accordingly, the global spatial autocorrelation (i.e., Moran's I statistics) analysis found significant positive Moran's I (i.e., I = 0.34, permutation 999, and p-value 0.001). The outcome is an indicative of the BLLs value's non- dispersions and non- randomness among children in the neighborhoods of Chicago city. On the other hand, the positive Moran's I coefficient suggests the spatial clustering of the similar values (i.e., high or low) of the BLLs values.

Furthermore, the Local Moran's statistic is conducted to analyze spatially-resolved locations where significant spatial clustering is present. Accordingly, the Local Moran's statistic found areas of significant clustering of higher values of the BLLs among children as well as low values (i.e., p-value = 0.05; See Figure 3c). Accordingly, neighborhoods in the south, far southwest, southwest and far south sides of Chicago have shown significant clustering of higher values of the BLLs among children (i.e., p-value = 0.05; Figure 3c). These sides of the city are areas known for their predominant residence of minority (i.e., Blacks and Hispanic American) population. On the other hand, neighborhoods in the central, north, and far north sides of Chicago have shown significant clustering of lower amounts of higher values of the BLLs among children (Figure 3c).

The geographical controls of the BLLs among children consists of studies conducted in Texas (Yu, 2016), in Greater Flint, Michigan (Hanna-Attisha et al., 2016), in Baltimore Maryland (Wheeler et al., 2019), and in Kabwe, Zambia (Moonga et al., 2021). It is also consistent with reported dynamics among various demographic groups in the state. In Illinois, although the state average is lower, the levels among African American and Hispanic children are reported to roughly two to three times the elevated BLLs among white children (Fokum et al., 2017).

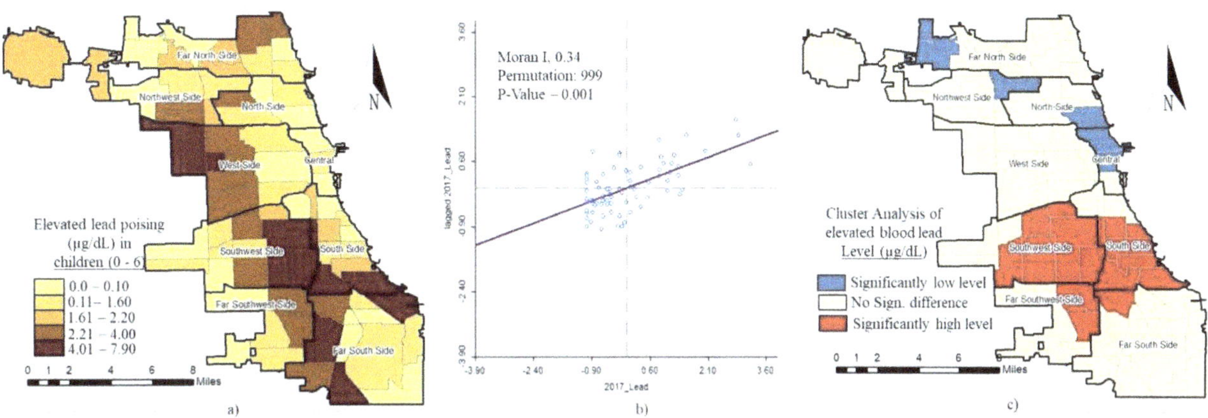


Figure 3: The 2017 data of elevated BLLs among the children aged 0 – 6 year among community areas of Chicago: a) spatial distribution; b) statistical evaluation of global spatial controls and c) statistical evaluation of local spatial controls.

## 4.2. Spatial relationship of elevated lead blood level and crime occurrence rate and test scores for the Chicago neighborhoods

The spatial associations were established between the BLLs, the crime occurrence rate, and students' ISAT test scores, (Figure 4 and Table 1). Accordingly, significant associations were found between the BLLs and average ISAT scores (i.e., P-value < 0.05) and crime occurrence rate (per 1000 residents) (i.e., P-value < 0.05). Additionally, with the $R^2$ of 40% and 53% for the average ISAT scores and crime occurrences rate, respectively, the associations were demonstrably strong. However, while the relation with the average ISAT scores is negative (B- coefficient = -0.08), the one with the crime occurrence rate is positive (B- coefficient = 0.007), indicating in the values of BLLs is associated with higher incidences of violent crimes and students' poor academic performances.

Table 1 Relationships of blood lead level among the children and ISAT score and Crime Occurrence Rate

| Variables | Strength ($R^2$) | Direction of the relationship (B- coefficient) | Local Variability (Sigma) | Significance of the relationship (P-value) |
|---|---|---|---|---|
| **Crime Occurrence Rate** | 0.53 | 0.007 | 1.103 | 0.000 |
| **Average ISAT Score** | 0.40 | -0.08 | 1.607 | 0.000 |

However, such relations are locally variable across the city (Figure 4). Accordingly, the local association (i.e., $R^2$) of the BLLs among children and crime occurrence rates (per 1000 residents) ranged between $R^2$ of 20% and 88% (i.e., moderate to very strong associations). Moderate associations ($R^2$ = 16% - 36%) observed in the neighborhoods of the north, far northern, and far southern sides of the city; whereas a very strong association ($R^2$ > 67%) were established in the southwest and far southwest sides.

Similarly, the local association (i.e., $R^2$) of the BLLs among children and averaged ISAT scores ranged between $R^2$ of 5% and 47% (i.e., weak to moderate associations). The weak association was observed in the neighborhood of Chatham, in the far south side of Chicago, while the moderate associations (i.e., $R^2$ = 16% - 36%) dominated the west and northwest sides. The types and significances of the associations found in this study were consistent with studies elsewhere (Zhang et al., 2013; Sauve-Syed, 2017; Boutwell et al., 2017; Winter & Sampson, 2017). For example, Zhang et al., 2013 established the association of elevated BLLs of children younger than 6 years old with poor academic achievement in Detroit, MI; whereas Boutwell et al. (2017) found that BLLs are statistically significant predictors of violent crime in St. Louis City, MO.

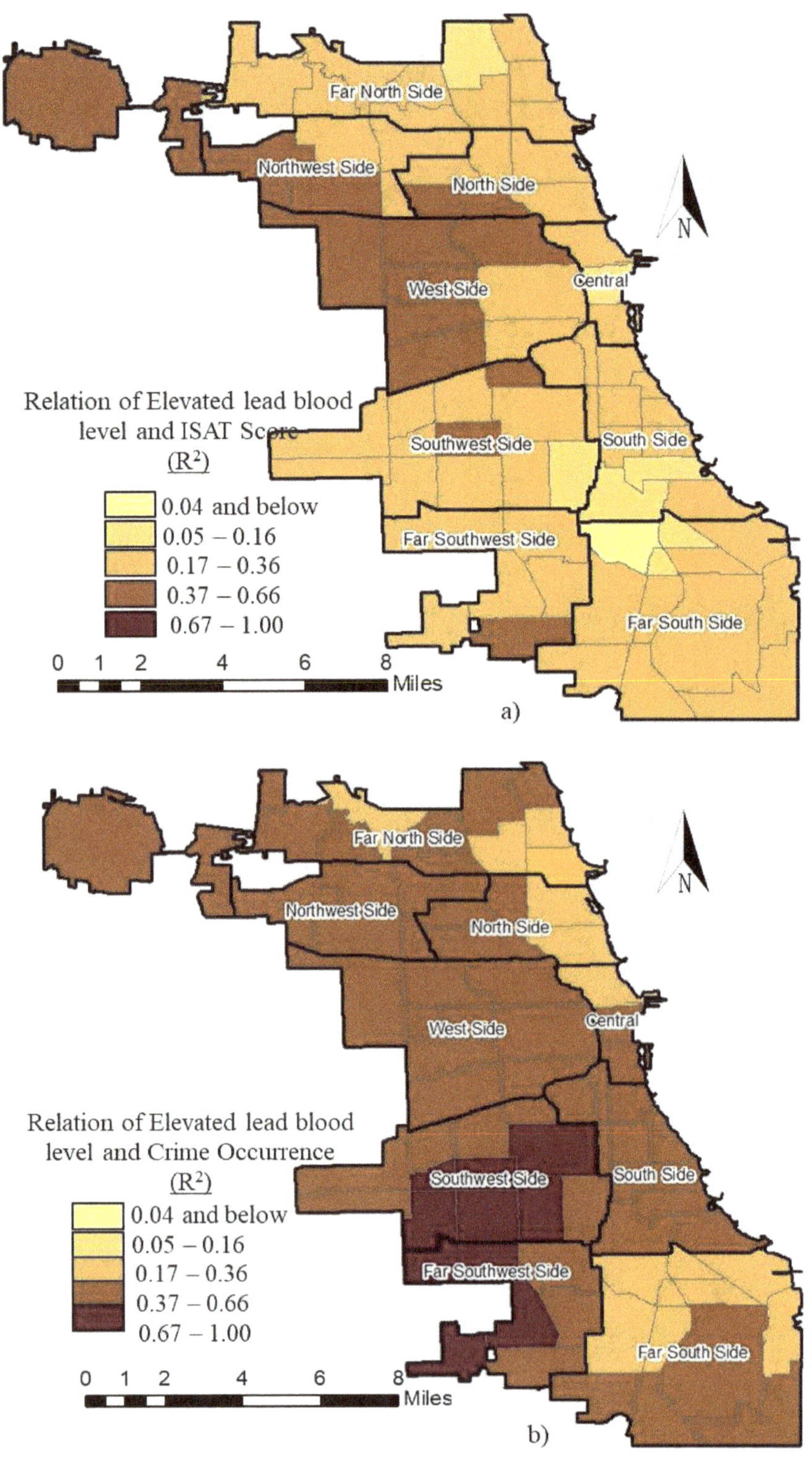



Figure 4: Geographically Weighted Spatial Relationships of elevated BLLs and the crime occurrence and students' test scores for the 77 community areas of Chicago

## 4.3. Analyzing the Spatial relationships of elevated lead blood levels and socioeconomic characteristics of the Chicago neighborhoods

Figure 5 investigated the controls of demographic and socio-economic variables on the BLLs among children. Accordingly, variables such as percent African American population, population below the poverty line, household's income, and the hardship index were considered. Statistically, significant associations of the BLLs and the above-mentioned variables were detected (i.e., P < 0.05), although the direction and the strength of the associations varies from variable to variable. The highest associations were found with percent population below the poverty line (i.e., $R^2$ = 54%), followed by the hardship index (i.e., $R^2$ = 47%; strong relation) and household per capital income (i.e., $R^2$ = 42%). The association of the BLLs and demographic variables, particularly percent African population, was relatively the least (i.e., $R^2$ = 40%). Similarly, while the relationships with percent African American population, below poverty lines, and the hardship index were positive, with household income, the relationship was negative. Low household incomes are associated with higher values of the BLLs among children in Chicago.

Table 2: Relationships of elevated BLLs among the children and socioeconomic characteristics

| Variables | Strength ($R^2$) | Direction of the relationship (B- coefficient) | Local Variability (Sigma) | Significance of the relationship (P-value) |
|---|---|---|---|---|
| Below poverty | 0.54 | 0.11 | 1.4 | 0.000 |
| Hardship Index | 0.50 | 0.04 | 1.4 | 0.000 |
| Percent African American | 0.40 | 0.02 | 1.6 | 0.000 |
| Per Capital Income | 0.42 | -0.0006 | 1.5 | 0.000 |

Figure 5 shows spatially-resolved strength of the associations of BLLs and selected demographic and socioeconomic variables using geographically weighted spatial regression analysis. For example, the strength of local association (i.e., $R^2$) of the BLLs among children and demographic characteristics (i.e., percent African American population) ranged between $R^2$ of 2% and 65% (i.e., very weak to strong associations). Generally strong associations ($R^2$ = 37% - 66%) were detected in the neighborhoods of the northwestern, west, southwest, and far southwestern sides of the city; while very weak and weak associations ($R^2$ = 2% - 16%) were detected in the central, south, and far south sides. On the other hand, local association (i.e., $R^2$) of socio-economic characteristics (i.e., aggregated household income) and the BLLs among children ranged between $R^2$ of 4% and 73% (i.e., very weak to very strong associations). The very strong associations ($R^2$ = 37% - 66%) were detected in the neighborhoods of the north and far north sides of the city, while very weak and weak associations ($R^2$ = 2% - 16%) were detected in the far south sides. The local associations of other socioeconomic variables (i.e., economic hardship index and percent population below poverty lines) were less variable (i.e., S.D. = 0.036 and 0.0001 respectively). In general, strong associations (i.e., 36% - 66%) were found throughout the neighborhoods of Chicago city.

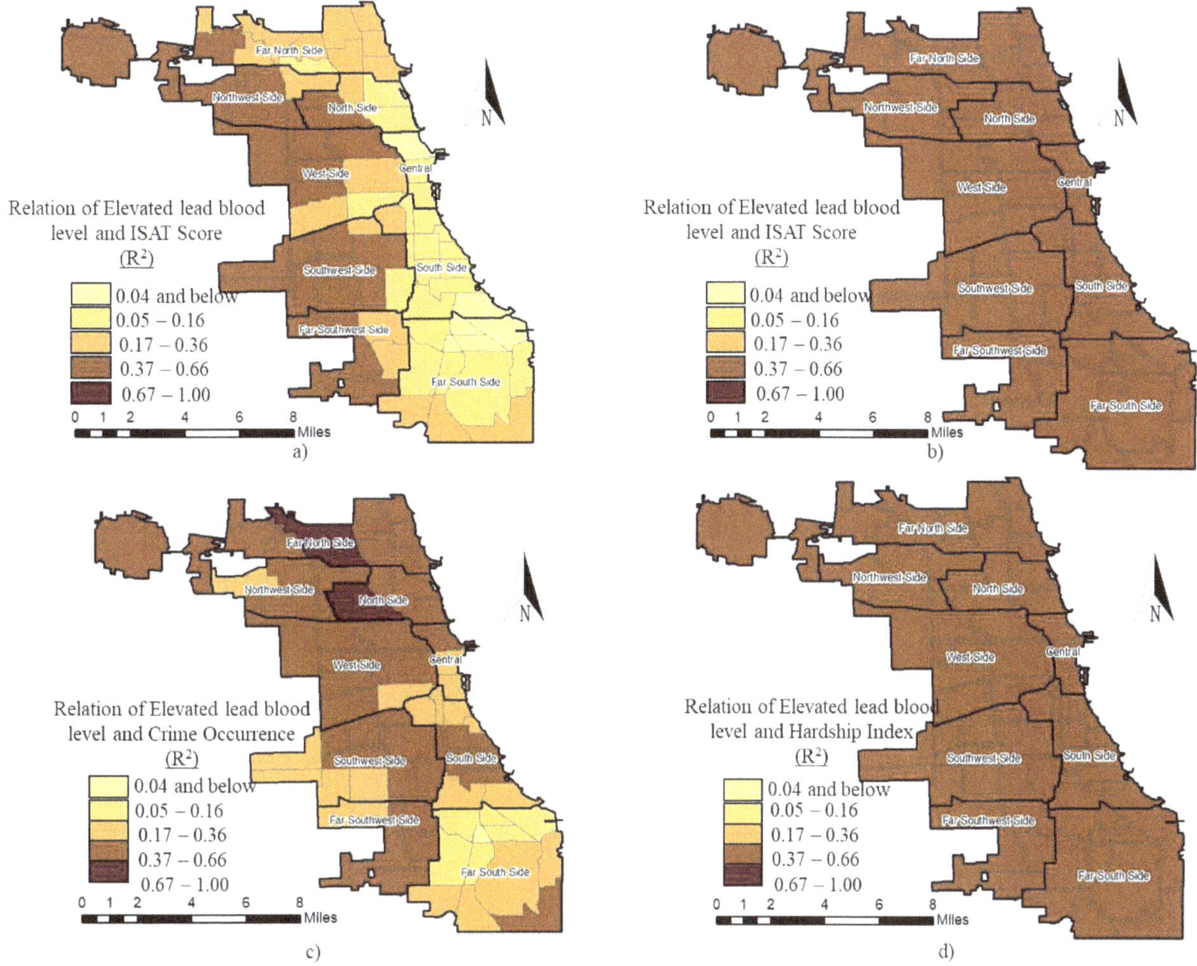


Figure 5: Geographically Weighted Spatial Relationships of elevated BLLs: a) percent African American population, b) economic hardship index, c) population's percent below poverty line, and d) household per Capital income of the community areas of Chicago.

The associations of BLLs and demographic and socioeconomic variables aggregated at the neighborhoods of Chicago city corroborates findings elsewhere (Morales et al., 2005; Sadler et al., 2017; Kim et al., 2018). Morales et al., 2005 reported the demographic and socioeconomic factors associated with the BLLs among Mexican-American children and adolescents in the United States, while Sadler et al., (2017) also documented BLLs associated with the oldest house age and poorest housing condition in Flint, MI. On the other hand, Kim et al., (2018) has shown a low socio-economic status associated with high BLLs among Korean children, although the underlying mechanism is not known.

## 5.1. Research Findings

This study aimed at evaluating the spatial distribution of children's elevated BLLs in Chicago's 77 community areas. Additionally, it was aimed at the spatial associations of BLL among children (i.e., 1 – 6 ages) and the social-economic, behavioral, and cognitive factors. The following research findings are reported.

1. Spatial patterns of the BLLs among children of Chicago are distributed in such a way that lower levels are found in the central, northern, far northern, and southwestern sides of the city, while the higher values are found in the western, southern, and southwestern sides of the city. Further analysis was conducted to see if there are significant overall geographical controls of the spatial patterns of the BLLs among children in Chicago. Accordingly, the global spatial autocorrelation (i.e., Moran's I statistics) analysis found significant positive Moran's I (i.e., I = 0.34, permutation 999, and p-value 0.001).

2. The associations of BLLs in children with students' learning performances and neighborhoods' record on criminal activities were established.  Significant associations were found between the Chicago neighborhoods' average ISAT performances and the BLL (i.e., P-value < 0.05). Similarly, significant association was also observed between neighborhood BLLs and the crime occurrence rate (per 1000 residents) and (i.e., P-value < 0.05). Higher strength of association was found with the crime occurrence rate (i.e., $R^2$ = 53%) vis-à-vis the average ISAT score (i.e., $R^2$ = 40%). However, while the association with the average ISAT scores is negative, the association with the crime occurrence rate is positive, indicating that higher values of

aggregated BLLs among children living in Chicago neighborhoods signifies the
likelihoods of higher criminal activity occurrences and poor students' performances in
learning.

3. Additionally, the association of the BLL among children and of Chicago neighborhoods'
demographic and socio-economic variables (i.e., percent African American population,
population below poverty line, household's income, and hardship index) were also
investigated.  Statistically, significant associations ranged between $R^2 = 40$ and $R^2 = 54$
were detected (i.e., $P < 0.05$). Similarly, while the associations with percent African
American, percent population below the poverty lines, and the hardship index were
positive, household income was negative; this indicates that neighborhoods with
aggregated higher BLLs imply the likelihood of poorer and blacker Chicago
neighborhoods.

## 5.2. Research Significance

This research is significant in assessing Children lead poisoning among Chicago
neighborhoods. As of 2012, the CDC lowered the testing standard BLLs of 10 micrograms
per deciliter to 5 micrograms per deciliter, and currently states there are no safe levels of
lead in blood.  Lead poisoning occurs when lead or any of its salts are absorbed into the
body.  Long-term exposure in young children could result in mental retardation, brain
damage, and behavioral problems such as criminal activities. Therefore, pediatricians
under the guidelines of the city of Chicago are mandated to follow-up with children up to
the age of 6 years old, and currently provide no other assistance beyond that age.

This study examined some major concerns surrounding its long-term effects and their relation with the social-economic, behavioral, and cognitive factors. The motivation for this focus is to review the number of children tested for and showing elevated BLLs with the number of reported crimes and schools with proven low-test scores. The dynamics of long-term lead poisoning of children can affect their lives well into adulthood; moreover, one of the effects could result in criminal behavior. There are significant correlations revealed by using spatial analyses between elevated blood lead poisoning with high crime areas, and low-test scores within the schools among low poverty communities of Chicago. Deteriorated paint containing lead in houses was believed to be the main pathway of lead exposure.  However, after 40 years of banning the use of lead-based paint, it continues to be a major issue in Chicago. Having such a sizable number of residents, the concern is to promote health, prevent disease, and reduce environmental hazards for all Chicagoans.

## 5.3. Research Limitations

Although, this research was significant in identifying the long-term effects of lead poisoning, it has the following limitations:

- Although the exploratory spatial data analysis toolsets in ArcGIS have been effective in investigating the geographical controls of elevated BLLs in Chicagoland, they also have some limitations. First, the hotspot areas do not necessarily mean the cluster of the highest values of elevated BLLs and, hence, can be misleading (Levine, 2013). The hotspots were, instead, indicating neighborhoods of the higher values relative to surrounding neighborhoods with the lower values. Secondly, both global and local Moran I have substantial type I errors since the significant test of spatial clustering is generally weak (Levine, 2013). The phenomenon is often allowing many neighborhoods in Chicagoland to show statistically significant hotspot areas. Reducing the search radius may reduce the size of type I errors and, thereby, the number of significant hotspot neighborhoods in Chicagoland; however, with irregular sizes of the neighborhood areas, a small search radius could make the community areas not have the surrounding neighborhoods to compare with.

☐ Again, even though the spatial regression models are effective tools in accounting for spatial dependency and nonstationary, while establishing relationships between elevated BLLs and social-economic, behavioral, and cognitive factors, they also have limitations. First, the algorithm of GWR is computationally intensive and, therefore, it takes a long time to run (Mitchell, 2012). Secondly, this tool, like other regression tools, is affected by the presence of multicollinearity in the data. Multicollinearity, which is common among spatially clustered explanatory variables, is the situation, in multivariate regression, where two or more involved explanatory variables are showing strong correlation and, hence, their values to the model are redundant (Wheeler, Tiefelsdorf, 2005).

☐ Tracking the children - City officials can mandate laws to alter health outcomes of children that have been affected by elevated BLLs. To this date the parents of the children that have tested positive for elevated BLLs do not receive any additional follow-ups or structural assistance with preventive measures beyond the age of five. These findings are based on children who have been tested for BLLs without tracking its cognitive and behavioral impacts. Such follow-up longitudinal studies could shed a better light in understanding locally applicable determinants or risk factors of the observed elevated BLLs among children in Chicago neighborhoods.

☐ This research depicted strong associations of the BLLs and the social-economic, behavioral, and cognitive characteristics of neighborhoods of Chicago city. However, the associations cannot be translated into causations such that the percent African American population, population below the poverty line, household's income, and the hardship index are considered responsible for elevated BLLs or vice versa. Further studies will be critical to establish how much of these associations are the causations.

# REFERENCES

Adams, D. A., Jajosky, R. A., Ajani, U., Kriseman, J., Sharp, P., Onwen, D. H., … & Abellera, J. P.
(2012). Centers for Disease Control and Prevention (CDC) 2014. Summary of
notifiable diseases—United States, 1-121.

Advisory Committee on Childhood Lead Poisoning Prevention (ACCLPP) (2012). Low Level
Lead Exposure Harms Children: A Renewed Call for Primary Prevention. Accessed
from: http://www.cdc.gov/nceh/lead/ACCLPP/Final_Document_030712.pdf.

Akkus, C., & Ozdenerol, E. (2014). Exploring childhood lead exposure through GIS: a review
of the recent literature. International Journal of Environmental Research and Public
Health, 11(6), 6314-6334.

Akkus, C. (2016). A Spatial Inquiry into Childhood Lead Poisoning in Shelby County,
Tennessee.

Anselin, L. (1995). Local indicators of spatial association—LISA. Geographical analysis,
27(2), 93-115.

Böhner, J., & Bechtel, B. (2018). GIS in climatology and meteorology. In Comprehensive
geographic information systems (pp. 196-235). Elsevier.

Britannica, T. Editors of Encyclopaedia (2020, January 9). *Lead. Encyclopedia Britannica*.
https://www.britannica.com/science/lead-chemical-element

Boutwell, B. B., Nelson, E. J., Qian, Z., Vaughn, M. G., Wright, J. P., Beaver, K. M., … &
Rosenfeld, R. (2017). Aggregate-level lead exposure, gun violence, homicide, and
rape. PloS one, 12(11), e0187953.

Clune, A. L., Falk, H., & Riederer, A. M. (2011). Mapping global environmental lead poisoning
in children. *Journal of Health and Pollution, 1*(2), 14-23.

Committee on Environmental Health. (2005). Lead exposure in children: prevention, detection, and management. Pediatrics, 116(4), 1036-1046.

Danger, K. I. (2008). The year of the recall–An examination of children's product recalls in 2007 and the implications for child safety.

Fokum, F. D., Shahidullah, M., Jorgensen, E., & Binns, H. (2017). Prevalence and Elimination of Childhood Lead Poisoning in Illinois, 1996–2012. In Applied Demography and Public Health in the 21st Century (pp. 221-236). Springer, Cham.

Geier, D. A., Kern, J. K., & Geier, M. R. (2017). Blood lead levels and learning disabilities: a cross-sectional study of the 2003–2004 National health and nutrition examination survey (NHANES). International journal of environmental research and public health, 14(10), 1202

Hanna-Attisha, M., LaChance, J., Sadler, R. C., & Champney Schnepp, A. (2016). Elevated blood lead levels in children associated with the Flint drinking water crisis: a spatial analysis of risk and public health response. American journal of public health, 106(2), 283-290.

Holmes, A. (2021). *Spatial Modeling of Naturally Occurring Retirement African Americans' Communities in Chicagoland* (Doctoral dissertation, Chicago State University).

Hunley III, R. (2019). The Great Migration? African American Population Growth And Decline In The Chicago Metropolitan Area (Doctoral dissertation, Chicago State University).

Hunley, R., & T. Gala, (2020). Spatial Patterns of African-American Population and Movement in Chicagoland. SSRG International Journal of Geoinformatics and Geological Science 7(2), 28-36.

Illinois Department of Public Health (2007) Children Enrolled in the Department of

    Healthcare and Family Services (HFS) Medical Programs Tested for Blood Lead

    Poisoning; State and Community Based, Illinois Lead Program, June 2007

Johnson, D. L., & Bretsch, J. K. (2002). Soil lead and children's blood lead levels in Syracuse,

    NY, USA. *Environmental Geochemistry and Health*, *24*(4), 375-385.

Jones, R. L., Homa, D. M., Meyer, P. A., Brody, D. J., Caldwell, K. L., Pirkle, J. L., & Brown, M. J.

    (2009). Trends in blood lead levels and blood lead testing among US children aged 1

    to 5 years, 1988–2004. Pediatrics, 123(3), e376-e385.

Knapp, A. (2013). How lead caused America's violent crime epidemic. Forbes. Com, 22.

Kim, E., Kwon, H. J., Ha, M., Lim, J., Lim, M. H., Yoo, S. J., & Paik, K. C. (2018). How does low

    socioeconomic status increase blood lead levels in Korean children?. International

    journal of environmental research and public health, 15(7), 1488.

Ko, S., Schaefer, P. D., Vicario, C. M., & Binns, H. J. (2007). Relationships of video

    assessments of touching and mouthing behaviors during outdoor play in urban

    residential yards to parental perceptions of child behaviors and blood lead

    levels. *Journal of exposure science & environmental epidemiology*, *17*(1), 47-57.

Kunene, N. R. (2016). Scaling up Spatiotemporal dynamics of HIV/AIDS Prevalence in sub-

    Saharan Africa (Doctoral dissertation, Chicago State University).

Kunene, N., Gella, M., & Gala, T. (2017). Geographic controls of adult HIV/AIDS prevalence

    and their determinants for Sub-Saharan Africa countries. *American Journal of Public

    Health*, *5*(4), 130-137.

Kunene, N., Ebomoyi, W., & Gala, T. S. (2018). Scaling-up spatiotemporal dynamics of

    HIV/AIDS prevalence rates of Sub-Saharan African countries. International Journal

    of Medical Engineering and Informatics, 10(1), 1-15.

Laidlaw, M. A., Zahran, S., Mielke, H. W., Taylor, M. P., & Filippelli, G. M. (2012). Re-

    suspension of lead contaminated urban soil as a dominant source of atmospheric

    lead in Birmingham, Chicago, Detroit and Pittsburgh, USA. *Atmospheric*

    *Environment*, *49*, 302-310.

Lanphear, B. P., Burgoon, D. A., Rust, S. W., Eberly, S., & Galke, W. (1998). Environmental

    exposures to lead and urban children's blood lead levels. Environmental Research,

    76(2), 120-130.

Lead Safe Illinois (2021) Ripple Effects of Childhood Lead Poisoning: Healthy Homes &

    Healthy Communities Initiative: Loyola University Chicago," Loyola University:

    Loyola University Chicago, accessed June 1, 2021,

    https://www.luc.edu/healthyhomes/leadsafeillinois/leadfacts/rippleeffectsofchild

    hoodleadpoisoning/

Levin, R., Brown, M. J., Kashtock, M. E., Jacobs, D. E., Whelan, E. A., Rodman, J., ... & Sinks, T.

    (2008). Lead exposures in US children, 2008: implications for

    prevention. *Environmental health perspectives*, *116*(10), 1285-1293.

Levine, N. (2013). Hot spot analysis of zones. *N Levine, CrimeStat: Spatial Statistics Program*

    *for the Analysis of Crime Incident Locations, Version*, *4*, 242960-242995.

Lenntech, (2018) Water Treatment and Purification. Lenntech. Retrieved from

    https://www.lenntech.com/periodic/elements/pb.htm#ixzz5EHutcM2p).

Mateo-Tomás, P., Olea, P. P., Jiménez-Moreno, M., Camarero, P. R., Sánchez-Barbudo, I. S., Rodríguez Martín-Doimeadios, R. C., & Mateo, R. (2016). Mapping the spatio-temporal risk of lead exposure in apex species for more effective mitigation. *Proceedings of the Royal Society B: Biological Sciences*, *283*(1835), 20160662.

Meyer, P. A., Pivetz, T., Dignam, T. A., Homa, D. M., Schoonover, J., Brody, D., & Centers for Disease Control and Prevention. (2003). Surveillance for elevated blood lead levels among children-United States, 1997-2001. Morbidity and Mortality Weekly Report CDC Surveillance Summaries, 52(10).

Mielke, H. W., Anderson, J. C., Berry, K. J., Mielke, P. W., Chaney, R. L., & Leech, M. (1983). Lead concentrations in inner-city soils as a factor in the child lead problem. *American Journal of Public Health*, *73*(12), 1366-1369.

Mielke, H. W. (1994). Lead in New Orleans soils: new images of an urban environment. *Environmental Geochemistry and Health*, *16*(3-4), 123-128.

Mielke, H. W., & Reagan, P. L. (1998). Soil is an important pathway of human lead exposure. Environmental health perspectives, 106(suppl 1), 217-229.

Mielke, H. W., Covington, T. P., Mielke Jr, P. W., Wolman, F. J., Powell, E. T., & Gonzales, C. R. (2011). Soil intervention as a strategy for lead exposure prevention: The New Orleans lead-safe childcare playground project. *Environmental Pollution*, *159*(8-9), 2071-2077.

Mielke, H. W., Gonzales, C. R., Powell, E. T., & Mielke, P. W. (2013). Environmental and health disparities in residential communities of New Orleans: The need for soil lead intervention to advance primary prevention. *Environment international*, *51*, 73-81.

Miranda, M. L., Anthopolos, R., & Hastings, D. (2011). A geospatial analysis of the effects of aviation gasoline on childhood blood lead levels. *Environmental Health Perspectives, 119*(10), 1513-1516.

Mitchell A. (2012). ESRI Guide to GIS Analysis, Volume 2: Spatial Measurements and Statistics. New York: ESRI Press.

Morales, L. S., Gutierrez, P., & Escarce, J. J. (2005). Demographic and socioeconomic factors associated with blood lead levels among Mexican-American children and adolescents in the United States. Public health reports, 120(4), 448-454.

Moonga, G., Chisola, M., Berger, U., Nowak, D., Yabe, J., Nakata, H., ... & Bose-O'Reilly, S. (2021). Geospatial approach to investigate spatial clustering and hotspots of blood lead levels in children within Kabwe, Zambia. medRxiv.

Nathan, R. P., & Adams, C. (1976). Understanding central city hardship. Political Science Quarterly, 91(1), 47-62.

Needleman, H. L., & Bellinger, D. (1991). The health effects of low-level exposure to lead. Annual review of public health, 12(1), 111-140.

Needleman, H. L., Riess, J. A., Tobin, M. J., Biesecker, G. E., & Greenhouse, J. B. (1996). Bone lead levels and delinquent behavior. Jama, 275(5), 363-369.

Occupational Safety and Health Administration (2007) Regulatory review 29 CFR 1926.62 Lead in Construction: Accessed from https://www.osha.gov/laws-regs/lookback/lead-construction-review

Peng, T., O'Connor, D., Zhao, B., Jin, Y., Zhang, Y., Tian, L., ... & Hou, D. (2019). Spatial distribution of lead contamination in soil and equipment dust at children's playgrounds in Beijing, China. *Environmental Pollution, 245*, 363-370.

Reed, H. (2018). Indiana's Public Health is in Jeopardy: Lessons to Learn from Toxic

Chemical Contamination in East Chicago. Ind. Health L. Rev., 15, 109.

Reissman, D. B., Staley, F., Curtis, G. B., & Kaufmann, R. B. (2001). Use of geographic

information system technology to aid Health Department decision making about

childhood lead poisoning prevention activities. *Environmental Health

Perspectives*, *109*(1), 89-94.

Sadler, R. C., LaChance, J., & Hanna-Attisha, M. (2017). Social and built environmental

correlates of predicted blood lead levels in the Flint water crisis. American journal

of public health, 107(5), 763-769.

Sample, A. Jennifer (2020) Childhood lead Poisoning: Management. UpToDate, last

modified May 6, 2020, accessed from

https://www.uptodate.com/contents/childhood-lead-poisoning-

management?topicRef=6493&source=see_link.

Sanders, T., Liu, Y., Buchner, V., & Tchounwou, P. B. (2009). Neurotoxic effects and

biomarkers of lead exposure: a review. *Reviews on environmental health*, *24*(1), 15.

Sauve-Syed, K. (2017). Lead exposure and student performance: A study of Flint schools.

Unpublished manuscript, Department of Economics, Syracuse University, Syracuse,

NY.

Stewart, L. R., Farver, J. R., Gorsevski, P. V., & Miner, J. G. (2014). Spatial prediction of blood

lead levels in children in Toledo, OH using fuzzy sets and the site-specific IEUBK

model. *Applied geochemistry*, *45*, 120-129.

Stretesky, P. B., & Lynch, M. J. (2004). The relationship between lead and crime. Journal of

Health and Social Behavior, 45(2), 214-229.

Switzer, C. S., & Bulan, L. A. (2002). CERCLA: comprehensive environmental response, compensation, and liability act (Superfund). American Bar Association.

Tarrago, O., & Brown, M. J. (2017). Case studies in environmental medicine (CSEM) lead toxicity. Agency for toxic substances and disease registry.

Yu, Q. (2016). Spatial statistical analysis of childhood blood lead exposure in Texas.

Vaidyanathan, A., Staley, F., Shire, J., Muthukumar, S., Kennedy, C., Meyer, P. A., & Brown, M. J. (2009). Screening for lead poisoning: a geospatial approach to determine testing of children in at-risk neighborhoods. *The Journal of pediatrics*, *154*(3), 409-414.

Vasovic, S. (2020). *Spatial Analysis of Unemployment Disparities in Chicago* (Doctoral dissertation, Chicago State University).

Wheeler, D., & Tiefelsdorf, M. (2005). Multicollinearity and correlation among local regression coefficients in geographically weighted regression. *Journal of Geographical Systems*, *7*(2), 161-187.

Wheeler, W., & Brown, M. J. (2013). Blood lead levels in children aged 1–5 years—United States, 1999–2010. MMWR. Morbidity and mortality weekly report, 62(13), 245.

Wheeler, D. C., Raman, S., Jones, R. M., Schootman, M., & Nelson, E. J. (2019). Bayesian deprivation index models for explaining variation in elevated blood lead levels among children in Maryland. Spatial and spatio-temporal epidemiology, 30, 100286.

Wilson, M; Tailor, A & Linares, A (2017) Chicago Community Area Economic Hardship Index (2017), Great Cities Institute, University of Illinois Chicago.

Winter, A. S., & Sampson, R. J. (2017). From lead exposure in early childhood to adolescent health: A Chicago birth cohort. American journal of public health, 107(9), 1496-1501.

Witzling, L., Wander, M., & Phillips, E. (2010). Testing and educating on urban soil lead: A case of Chicago community gardens. Journal of Agriculture, Food Systems, and Community Development, 1(2), 167-185.

World Health Organization. (2010). Childhood lead poisoning. June 9, 2021 Accessed from https://www.who.int/ceh/publications/leadguidance.pdf

Xiao, G., Hu, Y., Li, N., & Yang, D. (2018). Spatial autocorrelation analysis of monitoring data of heavy metals in rice in China. *Food Control*, *89*, 32-37.

Zhang, N., Baker, H. W., Tufts, M., Raymond, R. E., Salihu, H., & Elliott, M. R. (2013). Early childhood lead exposure and academic achievement: evidence from Detroit public schools, 2008–2010. American journal of public health, 103(3), e72-e77.

# About the Author

Deidre's interest in geography was first sparked after attending a townhall meeting as an assignment given by her Intro to Geography professor. Having always been intrigued by the physical features of earth, its atmosphere, and human activity, after that initial townhall, Deidre changed her major from Computer Science to Geography.

Deidre now holds both a bachelor's and master's degree in Geography with a concentration in Geographic Information Systems (GIS) from Chicago State University. She currently resides in the Greater Chicagoland area. For more information, contact Deidre White-Bradshaw at dbradshaw2025@gmail.com.